Áurea Nascimento de S. Mesquita
Tiago Fernando de Holanda
Elisabeth Regina Alves C. Silva

The issue of solid waste for basic education

Áurea Nascimento de S. Mesquita
Tiago Fernando de Holanda
Elisabeth Regina Alves C. Silva

The issue of solid waste for basic education

Weaving webs for critical environmental education

Imprint

Any brand names and product names mentioned in this book are subject to trademark, brand or patent protection and are trademarks or registered trademarks of their respective holders. The use of brand names, product names, common names, trade names, product descriptions etc. even without a particular marking in this work is in no way to be construed to mean that such names may be regarded as unrestricted in respect of trademark and brand protection legislation and could thus be used by anyone.

Cover image: www.ingimage.com

This book is a translation from the original published under ISBN 978-613-9-69979-7.

Publisher:
Sciencia Scripts
is a trademark of
Dodo Books Indian Ocean Ltd. and OmniScriptum S.R.L publishing group

120 High Road, East Finchley, London, N2 9ED, United Kingdom
Str. Armeneasca 28/1, office 1, Chisinau MD-2012, Republic of Moldova, Europe
Printed at: see last page
ISBN: 978-620-8-15076-1

I dedicate this work to my dear God, for the unconditional love he has for me.

ACKNOWLEDGEMENTS

I would like to thank God first of all for the opportunity to have started this wonderful course and today to be at the end of it, because without Him none of this would have been possible. To my dear parents Ugléces and Aurilene for their support and dedication, to my only brother Williams Siqueira for his help throughout the course, to my dear Aunt Edja, to my grandmother Marié (Vozita), to my grandmother Severina, to my beloved fiancé Gustavo Alexandre for his patience, support, understanding and love for everyone.

To my supervisor Dr Francisco Kennedy Silva dos Santos for his patience and opportunities during this time in academia.

To the teachers at the Joaquim Távora School, Professor Glauce, Professor Maria Clara, Professor Antônio, Professor Cida, Professor Enalva, Professor Elizabeth and the entire coordination team.

I would like to thank my dear teachers at the course, especially Professor Wagner Rocha for his teaching, his scholarships and his love of teaching the wonders of geography.

I would also like to thank my dear friends Aline França, Dayane Kelma, Adriana Silva and Smirna Albuquerque, for always believing in me, my friend Patrícia, her husband Júlio, my dear sister-in-law Simone, Istefany and all the friends and colleagues who are part of my story.

To my dear undergraduate teachers for their support, teaching and wonderful excursions, to my university friends: Daniele Cristina, Leandro Souza, Rafael Freitas, Eli and Jorge for their strength, help and for being together during these four years of many achievements, victories and lessons learnt. To the entire 2010.1 Night Geography class, it was great to meet you and be part of this class over the last two years. To our fellow PIBID students, our dear coordinator Professor Dr^3 Aldemir Dantas (in Memorian) and Professor Cláudio Ubiratan (Bira).

I would like to thank NEMA, especially Andrezza Karla, for her support and help with academic papers and articles. I would also like to thank the Geography Teaching and Teacher Professionalisation Research Group for their affection, events and constructive work during the academic year.

Finally, I would like to thank everyone who has helped me on this long academic journey and in the realisation of a dream that has only just begun...

"There is an appointed time for all things under heaven."
Ecclesiastes 3:1

"Many doors will open, Jesus will always be with you."

SUMMARY

This work is a set of reflections and analyses of the state of the art of the "waste issue" and how this theme is being addressed in basic education. We have taken Critical Environmental Education (CE) as our guiding principle, understood here as a methodology of analysis that starts from a sequential practice and apprehension of reality through perception and an attitude of critical-reflective investigation. We chose the issue of urban rubbish as our theme, as it is a fundamental issue to be worked on in the development of children and adolescents who are more concerned about ecological aspects such as rubbish in cities, air pollution, the issue of water, solid waste, among others. The central aim of this work is to map the state of the art on the issue of urban rubbish, with CAT as a fundamental category in the process of transforming the means and forms of apprehending reality. Based on this context, the question that drives this work has as its main locus the set of competences and skills to be mediated in the teaching-learning process based on the issue of rubbish.

Key words: Environmental Education, Solid Waste, Urban Waste

Summary

CHAPTER 1

INTRODUCTION

The term Environmental Education (EE) was used for the first time on record in 1948, in the city of Paris, at a meeting of the International Union for Conservation of Nature (IUCN). It wasn't until 1972, with the Stockholm Conference, that there was a greater influence on the international agenda in relation to environmental education. In 1975, in Belgrade (Yugoslavia), the International Environmental Education Programme was launched, which consequently defined the principles and guidelines for the future.

According to the Secretariat for Continuing Education, Literacy and Diversity/SECAD - MEC, in 1977, after a period of five (5) years, the Intergovernmental Conference on Environmental Education took place in Tbilisi (Georgia - former Soviet Union), chaired by the UN, which had recently helped to create the Environment Programme, as well as Unesco. At this meeting, which was signed by Brazil, definitions and objectives were drawn up, along with strategies for environmental education, which are currently being adopted around the world.

In 1992, another international document was presented, the Treaty on Environmental Education for Sustainable Societies and Global Responsibility, which was drawn up by global civil society in 1992 at the Global Forum during the United Nations Conference on Environment and Development (Rio 92). This document establishes fundamental principles of education for sustainable societies, as well as establishing links between public policies on environmental education and sustainability, providing an action plan for environmental education (MEC, 2007).

According to the National Curriculum Parameters of the High School - PCNEM - (Geography 1998, p.35) indicates as a way of obtaining students:

> Analyse and compare, on an interdisciplinary basis, the relationship between the preservation and degradation of life on the planet, with a view to understanding its dynamics and the globalisation of cultural, economic, technological and political phenomena that affect nature on different scales - local, regional, national and global.

According to the PCNEM for geography, we are looking for ways to encourage students to act sustainably in favour of the environment, making it possible to educate students who are critical and responsible for the survival of the planet. The Parâmetros Curriculares Nacionais do Meio Ambiente - PCN (1998, p.173) analyses that:

> The environmental perspective consists of a way of seeing the world in which the interrelationships and interdependence of the various elements in the constitution and maintenance of life are emphasised.

Urban waste is a major nuisance because it affects the quality of life of the world's population, from its production to its final destination.

One of modern society's biggest global problems is trying to reduce waste in urbanised areas. In addition to the significant increase in solid waste, especially in developing countries, there have been significant changes in its characteristics over the years. Nowadays, human beings have a strong connection with their environment, so much so that one of the main concerns of modern man is the environment.

The constant transformations in urban space have caused a number of environmental problems, due to the increase in population in cities, resulting in the need for housing and employment. These problems contribute to consumption and an increase in solid waste, leading to rubbish being dumped in prohibited places, such as water sources, canals and even on the streets.

According to the Brazilian standards ABNT NBR 10004:2004, Solid or Semi-Solid Waste is the result of human activities such as domestic, industrial, commercial, agricultural, service and sweeping. This includes sludge developed from water treatment systems that are generated by equipment installed with pollution control, as well as some liquids that are discharged into the sewage system or that require techniques that are feasible with advanced technologies.

Solid waste is an unwanted material, everyone has tried to get rid of it in some way, in other words, to dispose of it. It's a nuisance, because rubbish is irremediable, there's no way of avoiding it, because it's produced on a daily basis, and the whole production process generates waste and subsequently the end of the product's useful life.

Seeking to understand this socio-environmental problem, the aim of this paper is to carry out a theoretical and practical study of solid waste in urban centres and how we can reuse it in our daily lives, avoiding greater environmental impacts than those currently caused by various anthropogenic consequences.

The work focuses on recycling as a set of techniques that aim to reuse waste in the production cycle that it left. This activity results in the process of materials that were in the rubbish or would have been thrown away, being collected, separated and processed as raw materials for the construction of a new industrial or manual product. Recycling brings numerous benefits to urban centres, as reducing waste improves the cleanliness of cities, contributes to a better quality of life for the population, reduces pollution in the environment and contributes to the preservation of the natural resources that still exist.

The population living in urban areas today is proving to be one of the biggest problems on the planet. One of the major environmental consequences is the waste that is dumped untreated into the natural environment, which has consequently increased pollution of the air, water, soil and forests, with consequences for human health.

Raising awareness of the conscious disposal of waste in its many forms, especially in urban areas, emphasising special attention, given that the urban environment is an "urban ecosystem" and that most relationships are experienced in a fraction of nature, in this case "urban", which would affect not only the place in question, but in the medium and long term would affect nature.

CHAPTER 2

THE ISSUE OF URBAN WASTE THROUGH THE AGES

From the 19th century onwards, rubbish - solid waste between waste - faeces and urine - began to be defined and then collected separately through the sanitary sewer. Man has always separated himself from rubbish and waste, but it's not just humans who have this habit, animals also have this practice through the instinct to clean their nests and burrows. Man brings with him the instinct to clean in his work on urban cleaning (EIGENHEER, 2009).

As nomads, human beings didn't worry about the waste they generated, since they didn't stay in one place for long. The biggest problems arose from the moment villages began to appear, when man began to settle in a particular place (cities), which began around 4,000 BC. As well as understanding the use of irrigation, the Sumerians created complex cities with temples where they organised their supplies. Priests managed the city's cleanliness and water. The Babylonians built walled canals that interconnected for use in houses.

The Assyrians, who came after the Babylonians, channelled water and learnt to collect rainwater using the bricks of the time. Even smaller houses had a small drain to catch it. In the palace of King Sargon (2048-30 BC), there were toilets with seats and it is known that there was even running water to facilitate cleaning in these places. These practices were widely used by the Greeks through the Phoenicians.

Excavations at Harappa and Mohenjo-Daro, the main cities of the ancient Harappa culture, indicate that they were highly developed, including in terms of sanitary facilities. They also had paved streets, and in Mohenjo-Daro there were large underground lakes of various sizes with the possibility of inspection to collect wastewater and sewage through small drains coming from the houses.

In the case of the Egyptians, since 3000 B.C., they developed agricultural activities with flood irrigation through the River Nile, taking advantage of the river's waters. In the case of the Israelites, looking after rubbish and waste was of the utmost importance, because as nomads (going from one land to another), there was a rule of cleaning and maintaining the environment, based on biblical quotations from Deuteronomy 23:13-15 which says: "You must provide a place outside the camp for your needs. Along with your equipment, you must also have a shovel. When you go out to do your business, dig with it, and when you have finished, cover up the faeces. For Yahweh your God walks around the camp to protect you and to deliver your enemies from you. Therefore, your camp must be holy, lest Yahweh see something untoward in you and turn his back on you."

The places where the people of Israel made their worship sacrifices are also mentioned in Leviticus 4:11 and 12: 'The hide of the bullock and all its flesh, its head, its feet, its entrails and its dung, that is, the whole bullock, shall be taken outside the camp to a pure place, the place of the residue of the fat ashes. There he shall burn it on a wood fire; it is in the place of the residue of the fat ashes that the bullock shall be burned. And according to Leviticus 6:3 and 4, it says: "The priest shall put on his linen robe and cover his body with a linen breechcloth. Then he shall take the greasy ash from the burnt offering burnt by fire on the altar and put it beside the altar. Then he shall take off his garments, put on new ones and carry the greasy ash to a clean place outside the camp.

In Ancient Greece, the channelling and collection of water was already known, there were even toilets with running water to carry away faeces, and there was also the separation of water for general use (food, bathing) from toilets. Athens, around the 5th century BC, used a large amount of water for both domestic cleaning and bathing, so there was a greater need to channel water for supply and consumption.

Os romanos

Figura 1 Washing with Urine, Mural in Pompeii, 70s. Source: Lixo: Urban cleanliness through the ages, 2009.

The Roman cleaning system was built in the interests of the emperors when Rome reached its apogee. There are controversies about the size of the Roman population, but it is reasonable to say that in the early period of the Empire there were up to a million inhabitants. In Rome, the domus (private houses) of the patricians and the barracks for rent, in the 1st century BC the buildings were mostly vertical, reaching up to ten storeys, the property speculation of the time did not offer much comfort as it does today, the houses were in poor condition, with constant collapses and fires, the city even had a fire brigade with seven thousand men.

Rents were up to four times those of other Italian cities. Given the structure of the houses, the size and the climate of the city, water supply was one of the main needs, and in this area the Romans were famous. But the water that supplied the city had to come out, so the Romans became the major water collectors. It is in this context that the great urban and hygienic achievements of the Romans should be seen, and they are interesting when compared to today's large cities in undeveloped countries, with their marked contrasts.

Figura 2 Dumping waste in the Middle Ages - Engraving from 1489.
Source: Lixo: Urban cleanliness through the ages, 2009.

The collapse of the Roman Empire brought with it many health achievements, especially in Rome. Theodore the Great (493-526) tried to restore the Roman Empire's water systems and canals, but his successors did not follow, and the destruction and non-maintenance of these systems led to a number of disastrous health consequences. These consequences formed the diseases and epidemics that emerged. During the Pope's time in Avignon, Rome's population was reduced to 35,000 inhabitants. It was only later, with Frederick II (12121250), that care for these aspects was resumed. His health laws established rules for rubbish disposal and water supply (EIGENHEER, 2009).

In Italian cities during this period, regulations were established for the disposal of animal waste and for the keeping of animals within urban limits. An attempt was made to resume paving and the elimination of standing water. There was a ban on the improper disposal of waste by cart drivers, the dumping of rubbish and faeces in the streets and the use of rainwater as a means of getting rid of rubbish and waste, which caused canals to clog.

At the end of the Modern Age, not everything was idealised, but significant progress was made with the creation, in the 14th century, of public health initiatives to deal with sanitation and health issues in cities. This was the case in Konstanz in 1312 and Venice in 1485, which served as an example for other cities (EIGENHEER, 2009). Hoselassinala mentions that due to the dedication of doctors, it was possible to avoid various diseases and calamities in Germany in the 15th and 16th centuries.

In European cities in the middle of the Middle Ages, there were generally no streets, no pavements, no plumbing, no central water supply, no street lighting and no regular rubbish collection. Cologne in 1448. Brussels collected and composted its rubbish from 1560 onwards. Vienna started using carts in 1656, and in London in 1666, it is said that with an organised street cleaning service, citizens were drawn by lot who, on oath, took responsibility for the upkeep of areas of the city. At the time they were known as "scavengers", today they are called rubbish collectors, but the task was not willingly accepted, which caused the system to collapse. Innovations in urban cleaning therefore took place slowly in European cities, and in most of them without continuity.

In the Middle Ages, where cleaning services existed in urban areas, they were initially provided by private individuals, only when the work wasn't done properly did they opt for private individuals, there was also the help of prostitutes and prisoners who contributed to cleaning the city.

CHAPTER 3

URBAN WASTE IN BRAZIL

Figure 3 The garbage collector and the cart in 1930 in Rio de Janeiro. Source:
Lixo: Urban cleanliness through the ages, 2009.

In Brazil, to this day, there are profound regional, cultural and income distinctions. However, in their urban memory, few investments have been made in cleaning, which would serve as studies for a holistic and precise vision that would contribute to the country's question.

According to EIGENHEER (2009), Brazil faces immense difficulties when it comes to urban cleanliness. It is a continental country with uneven development. In view of this process, in order to get a more precise overview of the issue of rubbish, we focused on the city of Rio de Janeiro, looking for aspects that could contribute to understanding urban cleaning.

One of the concerns that affects the administration of Brazilian municipalities and the whole world is solid waste and its destination generated by various human activities. This waste, which can be liquid, gaseous or solid, can contaminate the environment when handled improperly, as well as polluting and causing waste to natural resources.

According to the Portal do Brasil website (Cf. http://www.brasil.gov.br/sobre/meio-ambiente/gestao-do-lixo), each individual in Brazil produces an average of 1.1 kilograms of rubbish a day. 188.8 tonnes of solid waste are constantly collected in Brazil, of which 50.8% is disposed of in inappropriate ways in the 2,906 dumps that Brazil already has. Around 27.7 per cent of cities dispose of solid waste in sanitary landfills and 22.5 per cent in controlled landfills, according to the National Basic Sanitation Survey by the Institute of Statistics.

According to the Federal Constitution, it is the responsibility of the municipal authorities to ensure cleanliness, rubbish collection, disposal and urban maintenance. With the National Solid Waste Policy law, municipalities have a solid foundation and a responsibility to change the potential of Brazil's waste landscape. According to the PNRS, the treatment and final destination of urban waste, such as industrial, hazardous, commercial and others, was approved in August 2010. Law 12305/2010 establishes the closure of rubbish dumps by 2014, prohibiting solid waste from being dumped in the open throughout the country and that only waste that cannot be recycled, after undergoing treatment and recovery, should be disposed of in environmentally appropriate locations.

In Brazil, according to a survey carried out by the Brazilian Institute of Geography and Statistics - IBGE, in 2010 rubbish collection increased in all Brazilian regions, the most widespread being in the Southeast with 95 per cent, followed by the South with 91.6% and the Centre-West Region with 89.7%, the lowest coverage was identified in the North and Northeast Regions, which showed an increase of 16.6 percentage points and 14.4 percentage points. In the urban regions, selective waste collection from households is above 90 per cent, ranging from 93.6 per cent in the North to 99.3 per cent in the South. In rural areas, Brazil increased its service compared to 2000, from 13.3 per cent to 26.9 per cent in 2010.

Waste disposal rose sharply in 2010, especially in rural areas. The obstacle and high cost of collecting rural rubbish has led residents to burn their rubbish to dispose of it. This type of alternative grew tenfold, rising to 48.2 per cent in 2000 and 58.1 per cent in 2010. To solve this problem, many residents dispose of their rubbish in vacant lots. In 2000, 20.8 per cent of residents adopted this alternative, but in 2010 it fell to 9.1 per cent, according to data from IBGE, 2010.

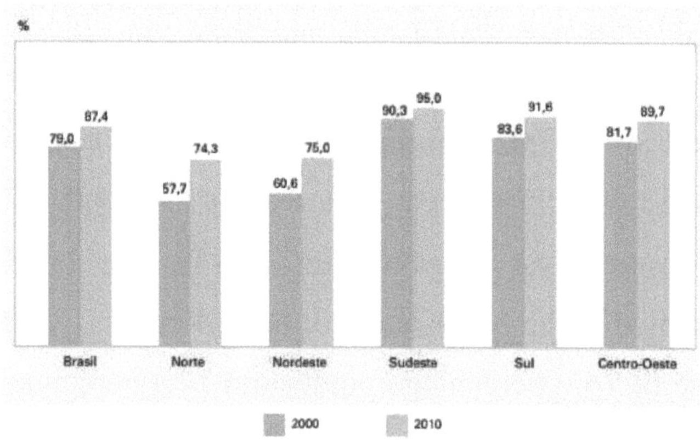

Figura 4 Proportion of permanent private households with rubbish collection, by major region - 2000/2010. Source:
IBGE, Demographic Census 2000/2010

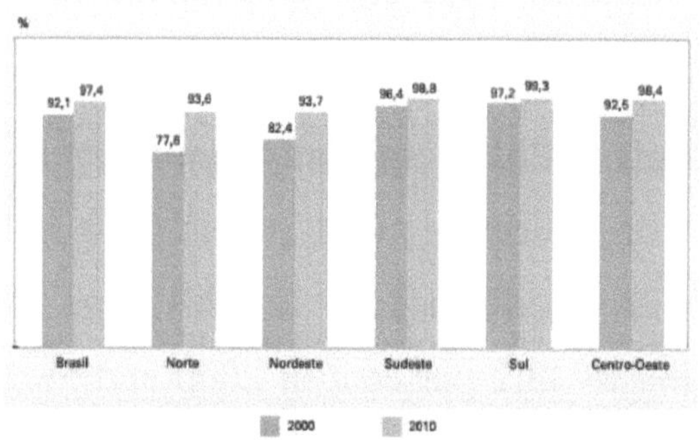

Figura 5 Proportion of urban permanent private households with rubbish collection, by Major Region - 2000/2010.
Source: IBGE, Demographic Census 2000/2010.

CHAPTER 4

SOLID WASTE IN URBAN AREAS

According to the Latin language, the word rubbish derives from Lix, which in Latin means ashes. Over a period of time, waste turns into ash from burning wood, but according to the Aurélio dictionary, rubbish is characterised as something useless, worthless, or even dirty. Based on the Brazilian Association of Technical Standards (ABNT), rubbish is defined as something that is undesirable or disposable and can be liquid, semi-solid (with a moisture content of less than 85%) or solid.

Solid waste is classified according to its type of origin (organic or inorganic). In order for solid waste to be properly characterised, its origin, constituents and characteristics must be known. During characterisation, which is carried out following specific sampling and testing standards, it is determined, for example, whether a waste is flammable, corrosive, combustible, toxic, etc. Its physical characteristics (granulometry, weight, volume, mechanical resistance, etc.) and chemical characteristics (reactivity, composition, solubility, etc.) are also studied.

Some of the standards used in this procedure are

- ABNT NBR10004/2007 - Solid Waste - Classification
- ABNT NBR10005:2004 - Procedure for obtaining leachate extract from solid waste

- ABNT NBR10006:2004 - Procedure for obtaining solubilised extract from solid waste
- ABNT NBR10007:2004 - Sampling solid waste
- ABNT NBR12808:1993 - Health Service Waste - Classification
- ABNT NBR14598:2000 - Petroleum products - Determination of flash point using the Pensky-Martens closed vessel apparatusUSEPA - SW846 - Test methods for evaluatingsolidwaste - Physical/chemicalmethod

4.1 Classification of solid waste

The purpose of classifying solid waste is to find out about the potential risks to the environment and public health so that it can be looked after properly. ABNT's NBR 10004:2004 classifies waste into three classes:

Class I waste - Hazardous;

Class II waste - Non-hazardous;

Class II A waste - Non-inert.

Class II B waste - Inert.

- **Class I - hazardous waste**

This type of waste requires special treatment and disposal due to its flammability, corrosiveness, reactivity, toxicity and pathogenicity. Examples: hospital waste, airport waste.

- **Class II waste - Non-hazardous**

Non-hazardous waste is waste that has a more appropriate destination for each classification, with the aim of reducing the impact on the environment. For example, uncontaminated restaurant food waste, cardboard waste, wood waste and sugar cane bagasse.

- **Class II A waste - Non-inert**

Non-inert waste is waste that is not hazardous but is not inert; it may have properties such as combustibility, biodegradability or solubility in water. They are domestic waste. Examples: sludge from the metallurgical industry, used tyres.

- **Class II B waste - Inert**

Inert waste is waste that comes into contact with water and remains drinkable. Most of this waste can be recycled, but it doesn't degrade very easily. Examples: demolition rubble, stones and sand removed from excavations.

CHAPTER 5

MAIN WAYS OF DISPOSING OF SOLID WASTE

There are countless ways of disposing of waste, such as landfill, composting, sterilisation, controlled landfill, industrial landfill, rubbish dumps. Although there are several ways of disposing of solid waste, most of them are unsuitable, as they can cause serious problems for human and environmental health, with the proliferation of diseases, pollution of the soil and the environment. The National Solid Waste Policy Law No. 12.305, of 2 August 2010, in its 6th article, states the prevention and precaution of waste, as well as sustainable development, with a measure of qualified services for environmental preservation, reducing impacts and the consumption of natural resources.

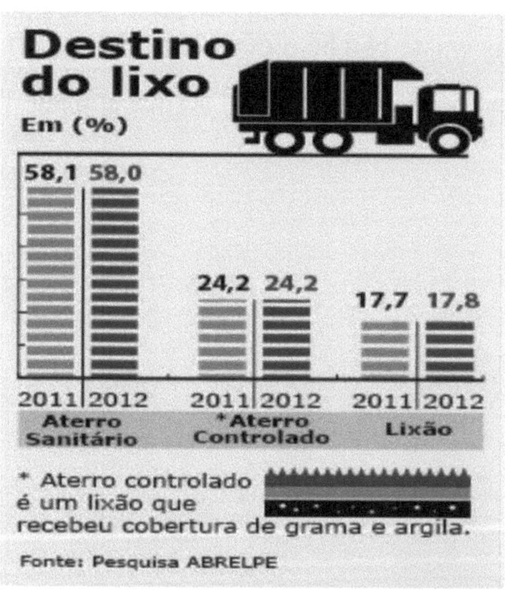

Figure 6 Destination of rubbish in Brazil. Source: Brazilian Association of Public Cleaning and Special Waste Companies (ABRELPE)

5.1 Open dump

Solid waste disposed of in open dumps can cause numerous public illnesses and damage the environment. In 2010, the Solid Waste Policy (Law 12.305/10) was approved by the National Congress, which requires Brazilian cities to put an end to rubbish dumps by August 2014. For this to happen, it will be necessary to organise selective rubbish collection and set up recycling plants, and organic materials will have to be deposited in controlled landfills.

According to the Brazilian Association of Public Cleaning Companies (Abrelpe), in 2012 around 58 per cent of solid waste in Brazil was improperly collected for landfill. The law stipulates in the Solid Waste Plans that municipal authorities are responsible for the management and collection of urban waste, as well as the selective collection and composting of organic waste. Society, in turn, must be users of these services established by the municipalities, and in the case of selective collection, they are responsible for separating solid, reusable and recyclable waste in an appropriate manner (Associação Brasileira de Empresas de Limpeza Pública e Resíduos Especiais, 2013).

Figure 7 Muribeca rubbish dump, Pernambuco. Source: Juliana Leitão, 2014

5.   2Disposal in a controlled landfill

The Controlled Landfill is a type of rubbish dump that, according to the legislation, allows waste to be disposed of in this particular location, but makes it clear that it is unsuitable for the environment, contaminating the natural soil.

According to NBR 8.419 ABNT (1985), it defines a controlled solid waste landfill in an urban environment as:

> Technique for disposing of solid urban waste on the ground without causing damage or risks to public health and safety, minimising environmental impacts. This method uses engineering principles to confine solid waste, covering it with a layer of inert material at the end of each working day.

The Controlled Landfill has no practices to combat pollution, due to the infiltration of rubbish into the soil and groundwater.

Generally, these types of landfills are old dumps that have undergone a process to reduce the effect of the leachate generated. This percolated liquid is often channelled for proper treatment, removing the gases that are produced at different depths of the landfill and covering the exposed cells on the surface, proper compaction, and managing the receipt of new waste.

5. 3Landfill

According to the Institute for Technological Research (ITP - 1995), a sanitary landfill is a process used to dispose of solid waste in the ground - particularly household waste - which, based on engineering criteria and specific operational standards, allows for safe confinement in terms of environmental pollution control and public health protection; consists of the technique of disposing of solid urban waste on the ground, without causing damage or risks to health and safety, minimising environmental impacts, a method which uses engineering principles to confine solid waste to the smallest possible area and reduce it to the smallest possible volume, covering it with a layer of earth at the end of each working day or shorter intervals if necessary (ABNT-1984)".

5. 4Composting

According to the MMA, composting is a process that is carried out by recycling organic matter that exists in vegetable or animal food waste. This process provides a useful destination for organic waste, avoiding its accumulation in landfills and improving the structure of soils, allowing a destination for organic waste, composting after this process is found as a final product, as: Organic Compost, which can be used in the soil without causing any harm to human health or the environment.

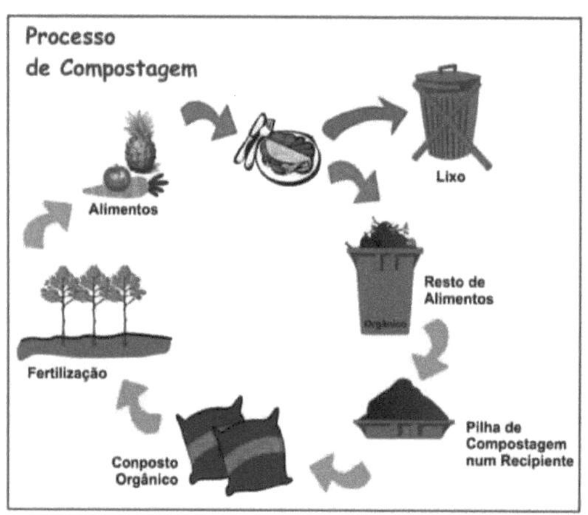

Figure 8 Composting process. Source: http://meioambiente.culturamix.com/blog/wp-content/gallery/1_21/the-types-of-composting-4.jpg

5.5 Bioremediation

They are disposed of in solid waste cells. After 1 to 2 years, these cells are excavated and all the material is sieved, then taken to separate the recyclables from the inert material, in which case what is left returns to the cell as covering material, as the same cell will have to be reused for new deposits.

5. 6Incinerator

This is a huge piece of equipment for burning healthcare or industrial waste (depending on the class). The waste is exposed to a high temperature (900°C on average). The organic compounds are reduced to their minimum constituents (carbon dioxide gas and water vapour) and the inorganic waste is transformed into ash. The advantage of the system is that it reduces the volume and neutralises the polluting action of the waste. This combustion takes place in an installation, usually called an incineration plant, designed and built for this purpose.

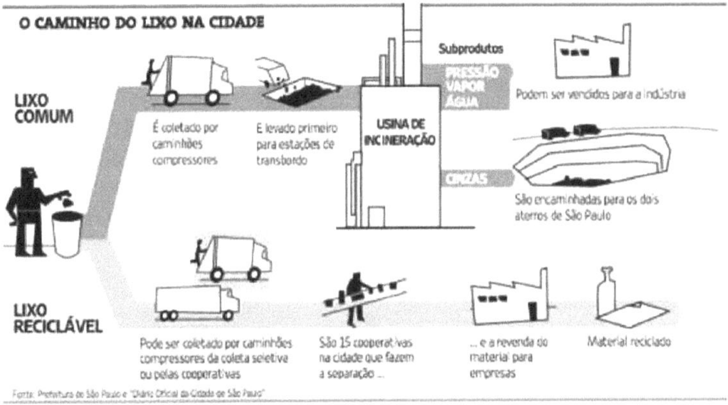

Figura 9   Rubbish collection - Incineration plant. Source: http://travessiambiental. blogspot.com. br/2011 /08/o-problema-do-rubbish-in-brazil-and-in-the-world.html.

CHAPTER 6

RECYCLING SOLID WASTE

According to the Ministry of the Environment (MMA), waste is recyclable, i.e. discarded waste that can be transformed and reused in parts or as a whole. Such materials can be reused in the production chain to vary the same product or use it in a different form to the original.

According to the World Health Organisation (WHO), solid waste is defined as any type of thing that an individual no longer wants, in a given place or at a given time, and that has no value. Europe, on the other hand, defines waste as "any substance or object which the holder discards or is obliged to discard by virtue of national provisions in force". (BIDONE, 2001).

6.1 Selective collection

In Brazil, around 443 municipalities have a selective collection programme, around 8% of the total. 22 million Brazilians have access to municipal selective collection programmes, although the number of cities providing this service has gradually increased, but most of them do not cover more than 10% of the local population, according to the Brazilian Institute of Geography and Statistics. The implementation of selective collection in the country is still in its infancy, as few municipalities have already implemented this type of collection.

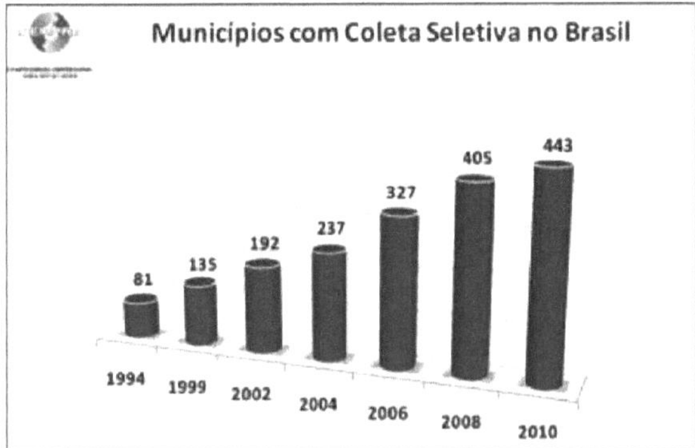

Figura 10   Figure 10: Register of Brazilian municipalities from 1994 to 2010. Source: http://www.cempre.org.br/ciclosoft 2010.php.

According to CONAMA resolution 275/2001, it establishes a colour code for the types of solid waste that exist, which is used to identify them when they are collected, transported and for information campaigns on selective collection.

Figura 11   Definition of collectors by colour for each type of material. Source: http://euacheiprimeiro.com/coleta seletiva

6.2 Valorisation of solid waste

The reuse of solid waste has been increasingly intensified, both in industrial production and in homes. Nowadays, the world is imposing the need for reuse, as natural resources are exhaustible, posing a risk to natural resources and the preservation of the planet.

In partnership with this objective, the planet has changed its habits and practices with the use of selective rubbish collection as a favoured alternative for urban waste treatment. As a result, plants of varying size and technology have been joined by other public and private initiatives involving industrial segments or sectors of the population that are especially differentiated (residential condominiums, commercial establishments, neighbourhoods, administrative regions and city halls), with the aim of reusing waste.

At the same time, large contingents of the poor population in Brazil's urban centres - the catadores, xepeiros - have an important strategy for survival in the rubbish trade.

6.3 The 3R's: Reduce, Reuse and Recycle

One of the main objectives of urban rubbish management in a city should be to reduce the amount of waste material that is disposed of directly. Reducing and reusing waste has brought many benefits to society, the economy and the environment. Solid waste management must be prioritised according to the 3R's concept **(Reduce, Reuse and Recycle).**

Reduce - Human beings must consciously learn to reduce rubbish and generate waste. From the moment this waste results in a burden for the public authorities and the taxpayer, reducing the volume of rubbish will mean reducing costs, as well as being a decisive factor in the preservation of natural resources. Less rubbish generated will also mean a smaller collection structure and lower final disposal costs.

Reuse - There are many ways to reuse objectives for financial or other reasons. Returnables and reusing disposable packaging for other purposes are just a few examples that can be used again.

Recycle - This is the transformation of the raw material into a form of recycling, now forming the third point in the **3 R's, and is** the alternative when it is no longer possible to reduce or utilise.

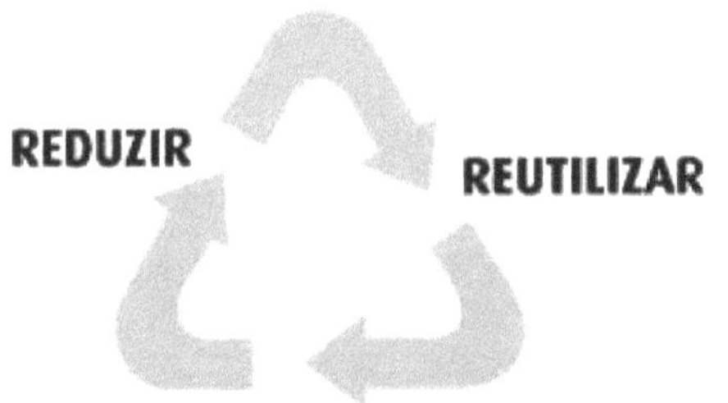

Figura 12 **3R's triangle model.**
Source:http://www.conscienciacoletiva.com.br/2011/04/regra-dos-3-rs.html

Figura 13   Solid Waste Management Scheme. Source: http://www.cchla.ufrn.br/geoesp/arquivos/sergio/TEXTOS/APOS TIL A.pdf

6.4 Recycling as social inclusion

Today in Brazil, there has been a lot of talk about separating rubbish, and people are starting to wake up to its social importance. This change contributes to the social inclusion of waste pickers. Currently, there are approximately 1 (one) million people in the country who walk the streets collecting recyclable materials with their carts.

Figura 14   Recyclable rubbish collector.
Source:http://www.henriqueafonso.com/noticias/henrique-afonso-wants-discuss-social-inclusion-of-waste-pickers/.

Figura 15   Sorting rubbish for recycling in industry.
Source:http://www.sobiologia.com.br/conteudos/reciclagem/reciclagem2.php

These waste pickers are essential for the recycling of these materials, and are divided into different categories, from small groups working on their own without protection or hygiene conditions to large companies working with machinery, vehicles and production control. Waste pickers who are self-employed are subjected to the exploitation of their materials, which they buy to resell for a minimal price for the hard work they do. According to the Business Commitment for Recycling (CEMPRE), only 10% of these waste pickers are organised and not dependent on intermediaries, as they work in recycling warehouses (fig. 08) that have adequate infrastructure for their work. In these co-operatives, the quantity of materials increases and the value increases, as the quality improves, making it an attractive price on the market and generating higher incomes and thus social gains.

CHAPTER 7

FINAL CONSIDERATIONS

From historical times to the present day, people have been thinking and reflecting on this issue, which, due to the expansion of the territory and urban growth with the development of industries and cities, has had disastrous consequences for humanity and the environment. The impacts of urban rubbish have raised social concerns that have been discussed in order to alleviate these problems, in search of concrete and economic solutions that contribute to a cleaner and more pleasant planet to live on.

This topic is of the utmost importance to students, as they begin to reflect on this issue, which is part of their daily lives, realising the importance of studying the environment in which they live.

The study of solid waste is therefore necessary for secondary school education, starting from the period of historical evolution in order to bridge the gap between the past and the present and interconnect what is happening today due to the consequences of the past. Students will learn about the basic concepts of solid waste, its stage, classification and the importance of reusing this waste, as these concepts are fundamental to their knowledge in order to make them reflect and from there they will begin to have their own opinions on the subject, making them opinion-makers based on what they have studied.

The topic makes students immerse themselves in the content more closely, as it is in their immediate reality, in their neighbourhoods, cities and even schools, to make them responsible citizens and aware of the actions taken in society, as solid waste surrounds space, making the whole of society participate from disposal to final collection.

CHAPTER 8

REFERENCES

BRAZILIAN ASSOCIATION OF PUBLIC CLEANING AND SPECIAL WASTE COMPANIES. ABRELPE Publications. **Solid Waste: Manual of Good Practices in Planning.** ABRELPE, São Paulo, 2013. Available at: <http://www.abrelpe.org.br/arquivos/manual_portugues_2013.pdf>. Accessed on: 18 Feb. 2014.

ABES - ASSOCIAÇÃO BRASILEIRA DE ENGENHARIA SANITARIA E AMBIENTAL.**Resíduos Sólidos Urbanos: Coleta e DestinoFinal**. Ceará, 2006. Available at: <http://www.cchla.ufrn.br/geoesp/arquivos/sergio/TEXTOS/APO STI LA.pdf>

BRAZILIAN TECHNICAL STANDARDS ASSOCIATION. NBR 8.419: **Presentation of Urban Solid Waste Landfill Projects:** procedures. Rio de Janeiro, 1985.

_____. **NBR 10004**: solid waste: classification. São Paulo, 2004.

NBR 8419 - presentation of urban waste landfill projects. São Paulo, 1984

BIDONE, Francisco Ricardo Andrade. **Solid Waste from Special Collections.** Rio de Janeiro: Rima2001.

BRAZIL. Law No. 12.305 of 2 August 2010. Establishes the National Solid Waste Policy; amends Law No. 9.605, of 12 February 1998; and makes other provisions. **Official Gazette of the Federative Republic of Brazil.** Brasília, DF, 2010. Available at: <http://www.planalto.gov.br/ccivil_03/_ato2007-2010/2010/lei/l12305.htm>. Accessed on: 20 Feb. 2014.

BRAZIL. Ministry of Education. Publications of the Secretary for Continuing Education, Literacy and Diversity. **Environmental Education: Sustainability Apprentices.** MEC, Brasília, 2007.

Available   at:
<http://portal.mec.gov.br/dmdocuments/publicacao2.pdf>.
Accessed on: 15 Feb. 2014.

BRAZIL. Ministry of the Environment. **Separate Waste Portal.**
Brasília, DF. Available at: <http://www.separeolixo.com/>.
Accessed on: 17 Feb. 2014.

______. Ministry of the Environment. **Composting**. Archive.
Available at:

<http://www.mma.gov.br/estruturas/secex_consumo/_arquivos/co
m post.pdf> Accessed on: 04 Mar. 2014.

BRAZIL. Resolution No. 275 of the National Environment
Council, of 25 April 2001. Establishes the colour code for the
different types of waste, to be adopted in the identification of
collectors and transporters, as well as in information campaigns for
selective collection. **Official Gazette of the Federative Republic of
Brazil.** Brasília, DF, 2001. Available at:
<http://www.mma.gov.br/port/conama/legiabre.cfm?codlegi=273
>.
Accessed on: 20 Feb. 2014.

CORPORATE COMMITMENT TO RECYCLING. CEMPRE's
collection of articles on the National Solid Waste Policy. **The
impacts of the new law against global warming.** Available at
<http://www.cempre.org.br/download/pnrs_001.pdf>. Accessed
on: 15 Feb. 2014.

EIGENHEER, Emílio Maciel.**Lixo: A limpeza urbana através dos
tempos**. Porto Alegre - RS, 2009.

HAJE, L. Solid Waste Policy foresees the end of rubbish dumps by
2014. **News portal of the Chamber of Deputies/Environment**,
Brasília, DF, 12 July 2013. Available
at:<http://www2.camara.leg.br/camaranoticias/noticias/MEIO-
AMBIENTE/447523-POLITICA-DE-RESIDUOS-SOLIDOS-
PREVE- O-FIM-DOS-LIXOES-ATE-2014.html>. Accessed on:
20 Feb. 2014.

BRAZILIAN INSTITUTE OF GEOGRAPHY AND STATISTICS. IBGE publications on Natural Resources. **Sustainable Development Indicators.** IBGE, Rio de Janeiro, 2012. Available at   : <http://www.ibge.gov.br/home/geociencias/recursosnaturais/ids/id s2 010.pdf>. Accessed on: 17 Feb. 2014.

INSTITUTO DE PESQUISAS TECNOLÓGICAS - IPT & COMPROMISSO EMPRESARIAL PARA RECICLAGEM - CEMPRE. **Municipal waste: integrated management manual.** IPT-CEMPRE, São Paulo, 1995.

LEITÃO, J. A soap opera called Lixão da Muribeca. **Diário de Pernambuco website.** Recife, 2010. Available at:

<http://www.old.pernambuco.eom/diario/especiais/meio_ambiente /n ovela.shtm>. Accessed on: 14 Feb. 2014.

MENDONÇA, Rita. Environmental educators teach through their attitudes (2012). **Revista   NovaEscola,** Available at: http://revistaescola.abril.com.br/ciencias/fundamentos/rita-mendonca-educador-ambiental-ensina-suas-atitudes-426107.shtml (accessed 07/09/2012).

PCNEM - National High School Curriculum Parameters for Geography. **Ministry of Education website.** Available at: < http://portal.mec.gov.br/seb/arquivos/pdf/cienciah.pdf>. Accessed on: 10 February 2014.

PCN - National Curriculum Parameters for the Environment. **Ministry of Education Portal.** Available at: <http://portal.mec.gov.br/seb/arquivos/pdf/meioambiente.pdf>. Accessed on: 12 Feb. 2014.

Printed by Books on Demand GmbH, Norderstedt / Germany